EXPLORING TRIANGLES

Paper-Folding Geometry

BY JO PHILLIPS

ILLUSTRATED BY JIM ROLLING

EXPLORING TRIANGLES

Paper-Folding Geometry

THOMAS Y. CROWELL COMPANY • NEW YORK

YOUNG MATH BOOKS

Edited by Dr. Max Beberman, Director of the Committee on School Mathematics Projects, University of Illinois

BIGGER AND SMALLER
CIRCLES
COMPUTERS
THE ELLIPSE
ESTIMATION
FRACTIONS ARE PARTS OF THINGS
GRAPH GAMES
LINES, SEGMENTS, POLYGONS
LONG, SHORT, HIGH, LOW, THIN, WIDE
MATHEMATICAL GAMES FOR ONE OR TWO
ODDS AND EVENS
PROBABILITY
RIGHT ANGLES: PAPER-FOLDING GEOMETRY
RUBBER BANDS, BASEBALLS AND DOUGHNUTS: A BOOK ABOUT TOPOLOGY
STRAIGHT LINES, PARALLEL LINES, PERPENDICULAR LINES
WEIGHING & BALANCING
WHAT IS SYMMETRY?

Edited by Dorothy Bloomfield, Mathematics Specialist, Bank Street College of Education

AREA
AVERAGES
BASE FIVE
BUILDING TABLES ON TABLES: A Book About Multiplication
EXPLORING TRIANGLES: Paper-Folding Geometry
A GAME OF FUNCTIONS
LESS THAN NOTHING IS REALLY SOMETHING
MAPS, TRACKS, AND THE BRIDGES OF KONIGSBERG: A Book About Networks
MEASURE WITH METRIC
NUMBER IDEAS THROUGH PICTURES
SHADOW GEOMETRY
SPIRALS
STATISTICS
3D, 2D, 1D
VENN DIAGRAMS

 Published simultaneously in Canada by Fitzhenry & Whiteside Limited, Toronto. Manufactured in the United States of America.

Library of Congress Cataloging in Publication Data. Phillips, Jo McKeeby. Exploring triangles: paper-folding geometry. SUMMARY: Basic concepts of triangles are revealed through paper-folding activities. 1. Triangle—Juv. lit. 2. Paper work—Juv. lit. [1. Triangle. 2. Geometry. 3. Paper work] I. Rolling, James H., illus. II. Title. QA482.P45 516'.22 74-14862 ISBN 0-690-00644-6 ISBN 0-690-00645-4 (lib. bdg.)

1 2 3 4 5 6 7 8 9 10

EXPLORING TRIANGLES
Paper-Folding Geometry

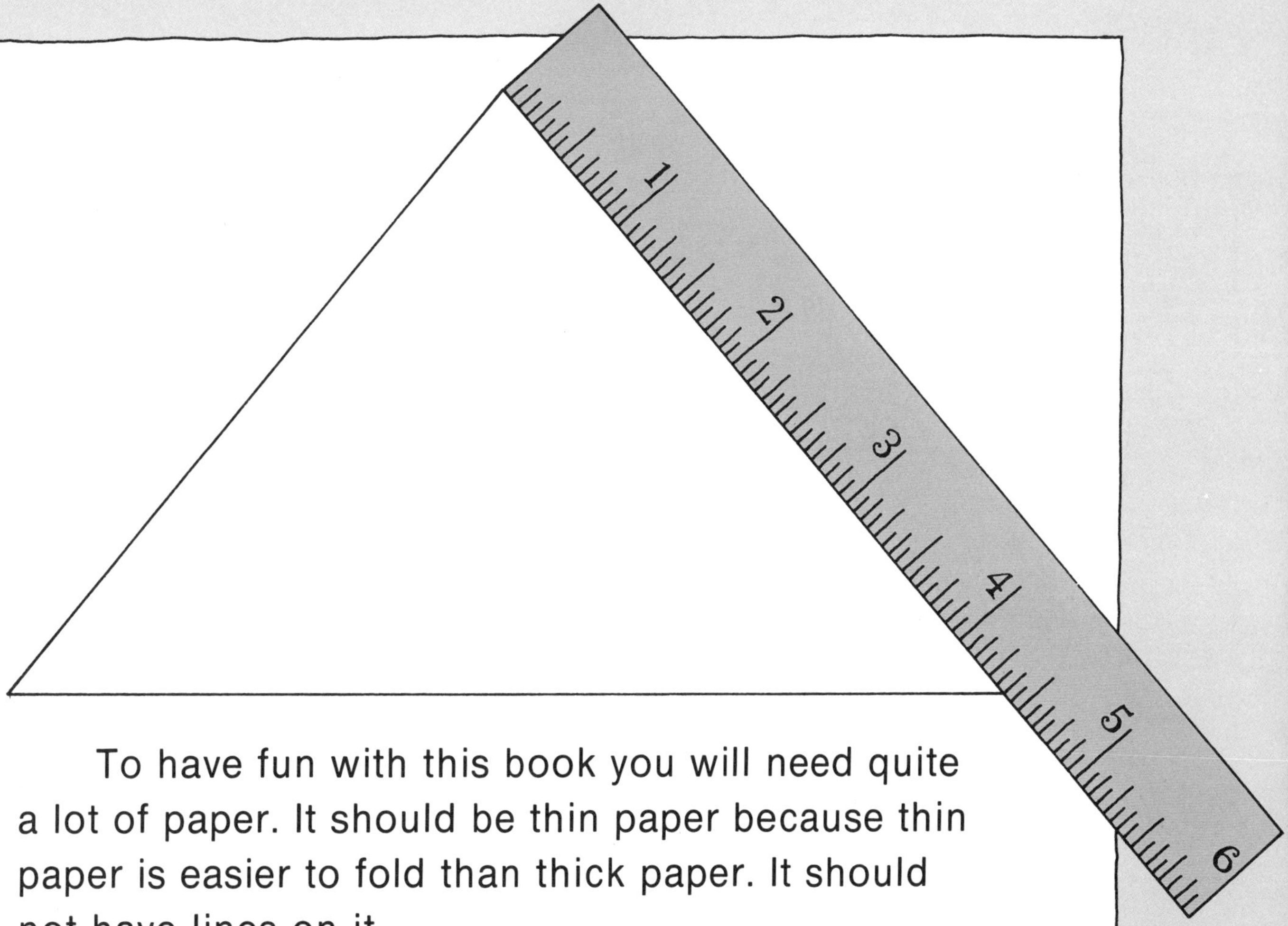

To have fun with this book you will need quite a lot of paper. It should be thin paper because thin paper is easier to fold than thick paper. It should not have lines on it.

You will need scissors, too, and a pencil. You also need a large card or a ruler to help you draw straight lines, and a compass for drawing circles.

Get your things together and draw a triangle like this one.

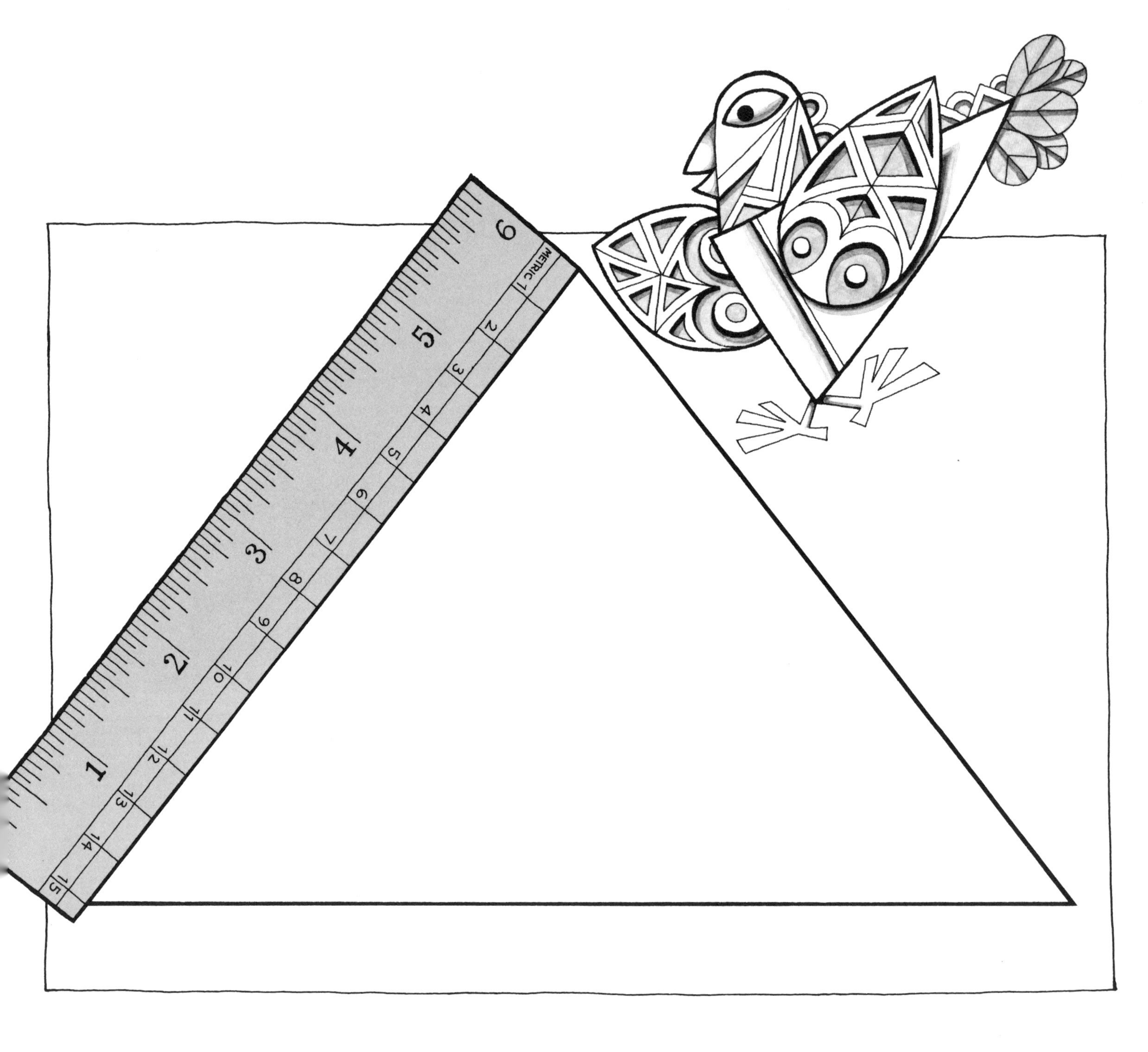
1
2
3
4
5
6
METRIC 1
2
3
4
5
6
7
8
9
10
11
12
13
14
15

The important thing about your triangle is that it should have two sides alike and the third side different.

Start with the two sides that are the same length. If you are using an inch ruler, make them each 6 inches long. If you are using a centimeter ruler, make them each 15 centimeters long. If you are using a card, make them the length of one side of that card. Have them come together almost, but not quite, in a square corner (a square corner is a RIGHT ANGLE).

Connect their other ends to make the third side of your triangle. The third side should turn out to be a little longer than the sides you measured.

Practice drawing, if you need to, until you get a really good copy of the triangle.

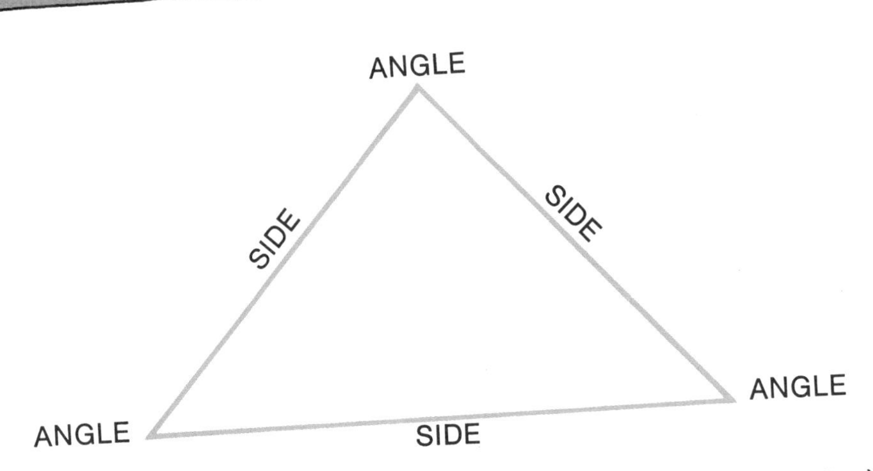

TRI means THREE. (A tricycle has three wheels.)
A triangle has three angles.
A triangle has three sides.
Did you draw three sides?
Did that give you three angles?

If you made a good drawing of this triangle, you can find a way to fold it so that two parts just match. Do that. Hold it up to the light so you can tell.

Make another copy of the same triangle. Cut this one out. Be careful to cut on the lines. Fold this cutout triangle so that the two parts just match.

Is it easier for you to fold your triangle on the whole sheet of paper or with the triangle cut out? Use either way you like for other experiments. You may wish to use one way for some experiments and the other way for others.

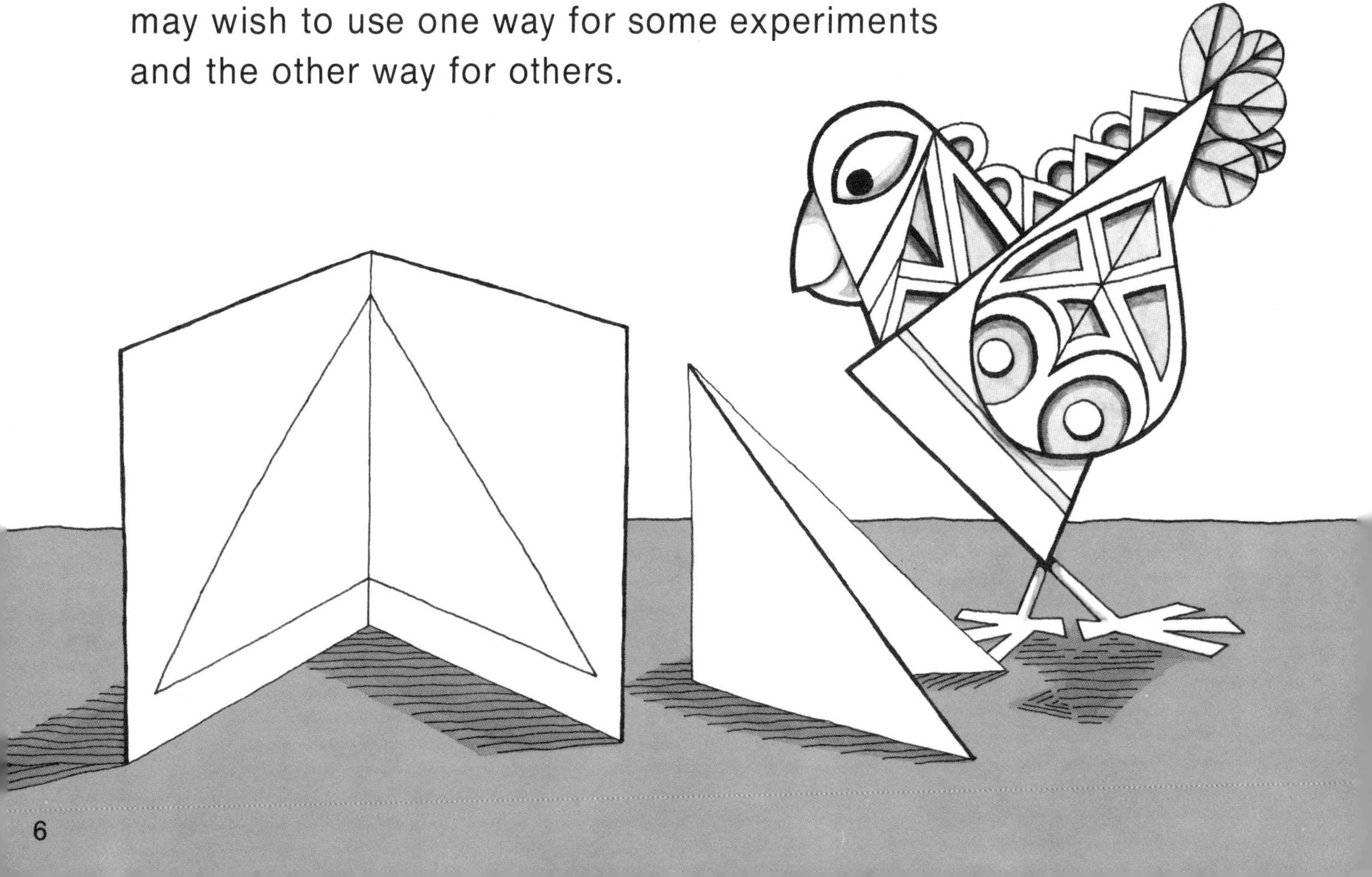

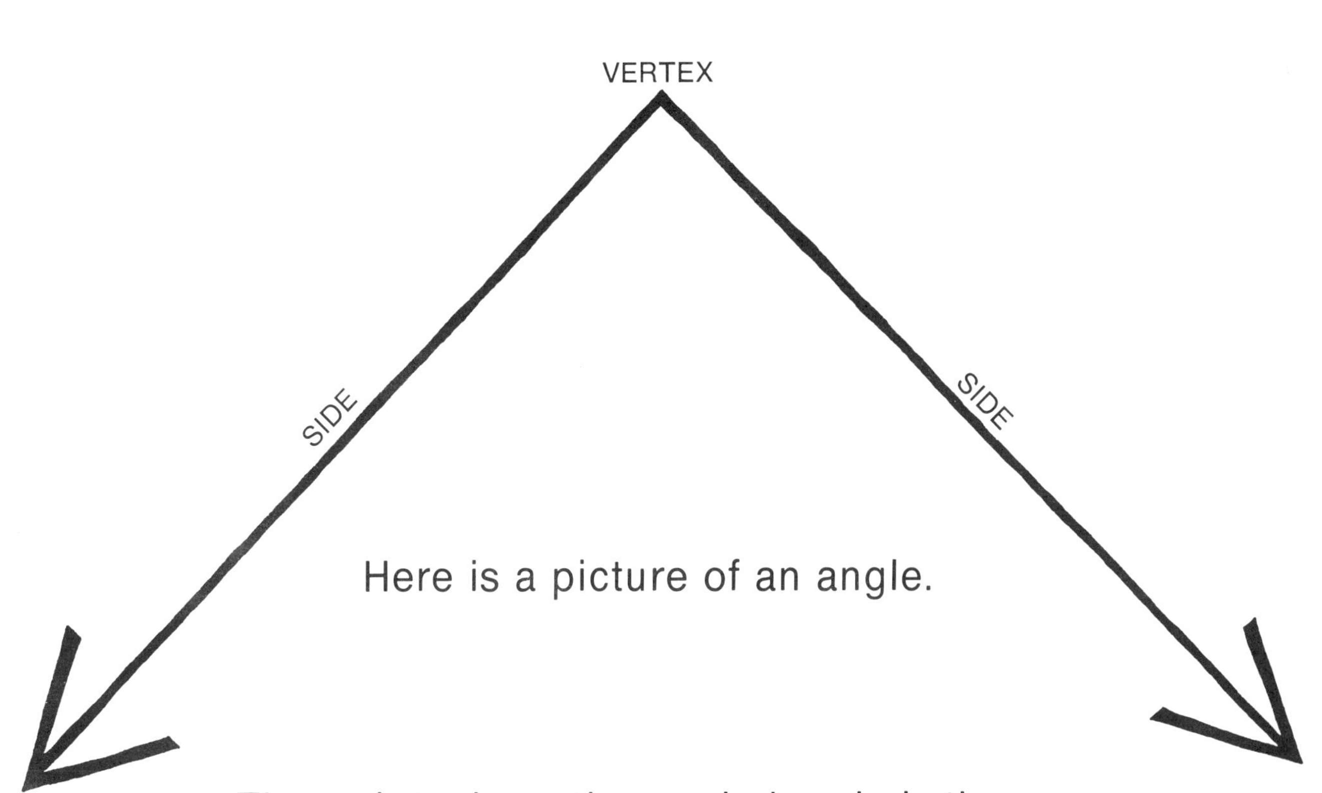

Here is a picture of an angle.

The point where the angle bends is the VERTEX of the angle.

The straight parts are sides. The sides of an angle start at the vertex.

The arrows in the picture mean that the sides do not have a stopping point. The sides of an angle go on and on.

The sides of an angle are RAYS. A ray has a starting point, but no stopping point. We say a ray has one ENDPOINT. When rays form an angle, the endpoint of each ray is the vertex of that angle.

Here are some more pictures of angles. Draw angles that look like these.

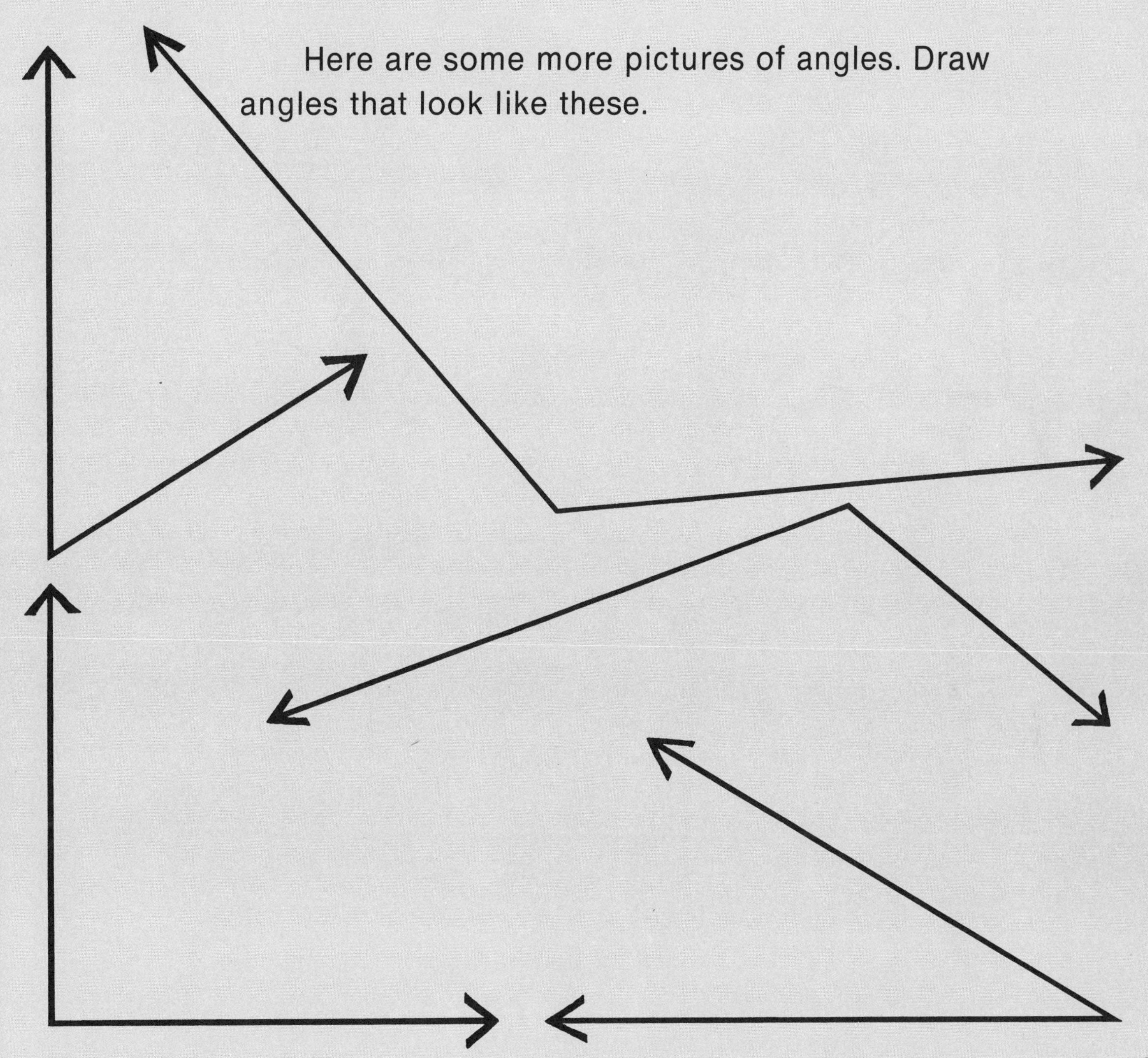

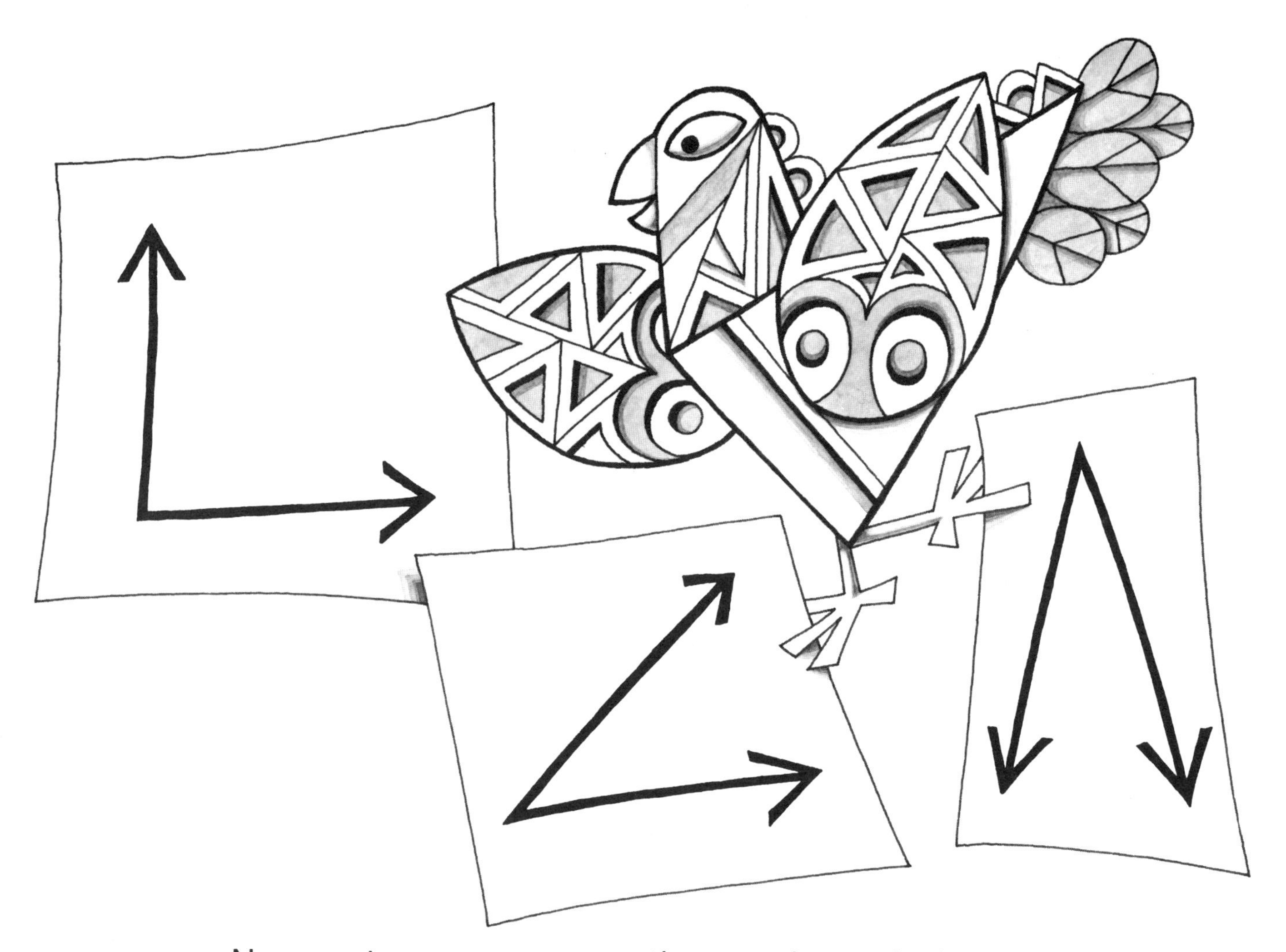

Now cut your paper so that each angle is on its own separate piece.

You can always fold a tracing of an angle so that the two parts match exactly. When you do this, the line of the fold BISECTS the angle.

To bisect anything is to divide it into two parts that are the same size. The line that bisects an angle is the BISECTOR of the angle.

Bisect each of your angles by folding it. Be careful to make each fold go through the vertex of the angle you are bisecting.

How many bisectors does an angle have?

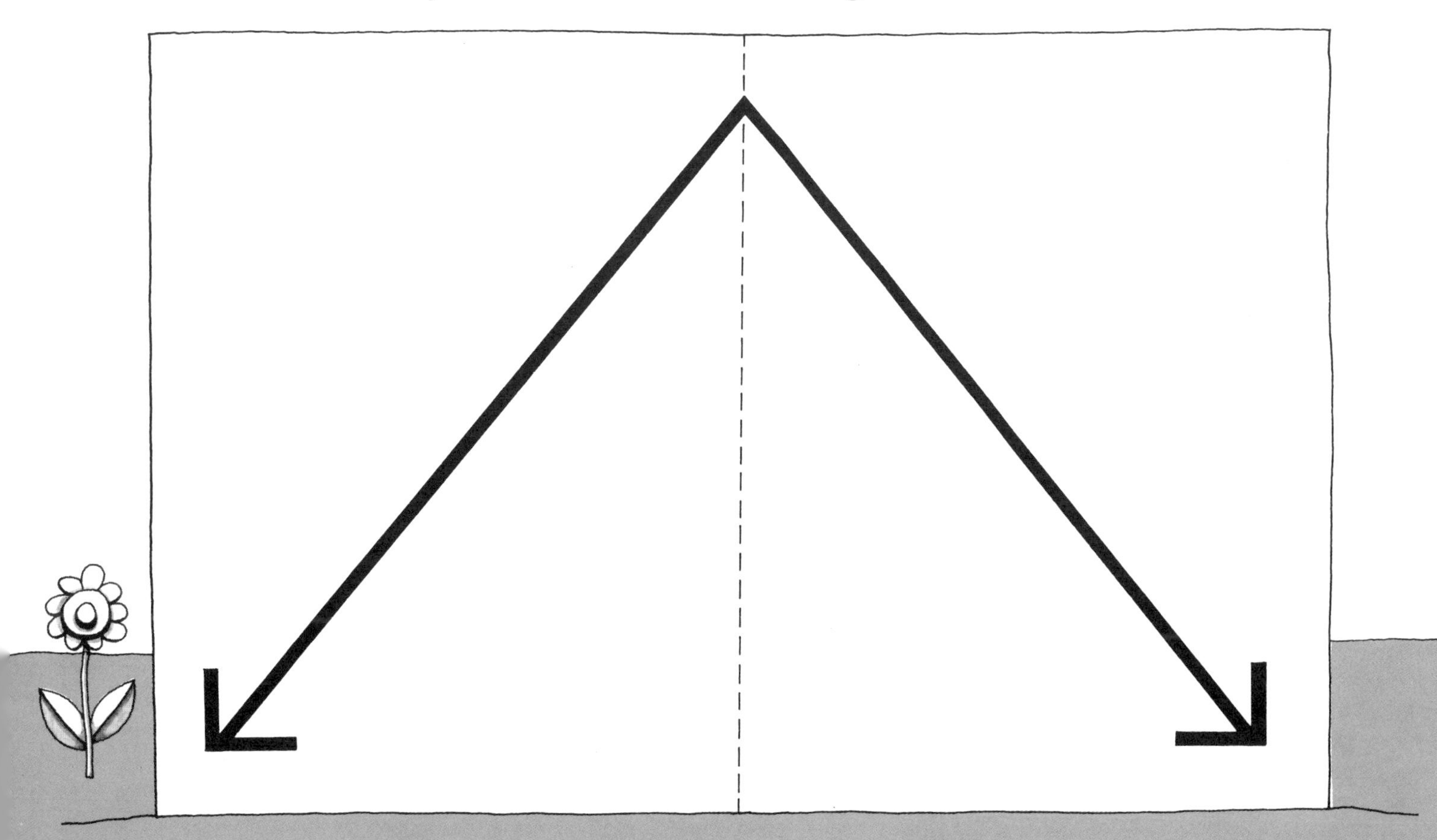

Now take one of the triangles you worked on before. Does the fold you made in it bisect one of the angles?

Bisect the other two angles by folding them. Be sure each bisector goes through a vertex.

You should find that the three angle bisectors of this triangle all go through the same point.

Do you think that is true only of triangles shaped like this one? Try it with some other triangles.

Copy these triangles that have different shapes. Maybe you can think of some more. Fold three angle bisectors for each of them.

Did the angle bisectors meet at one point each time? Do you think this would be true of all triangles? If you do, you are correct. Mathematicians can prove that this is true.

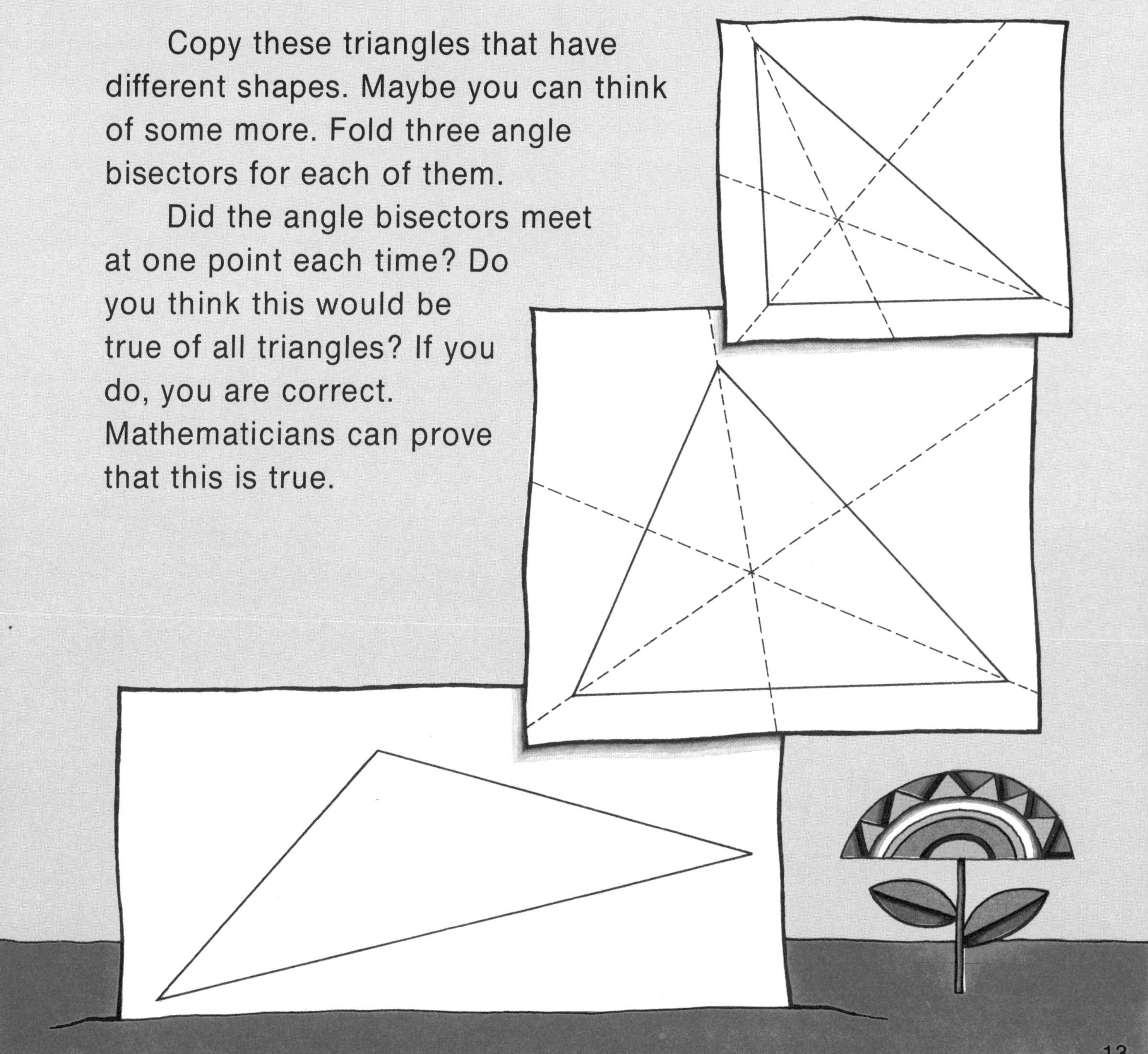

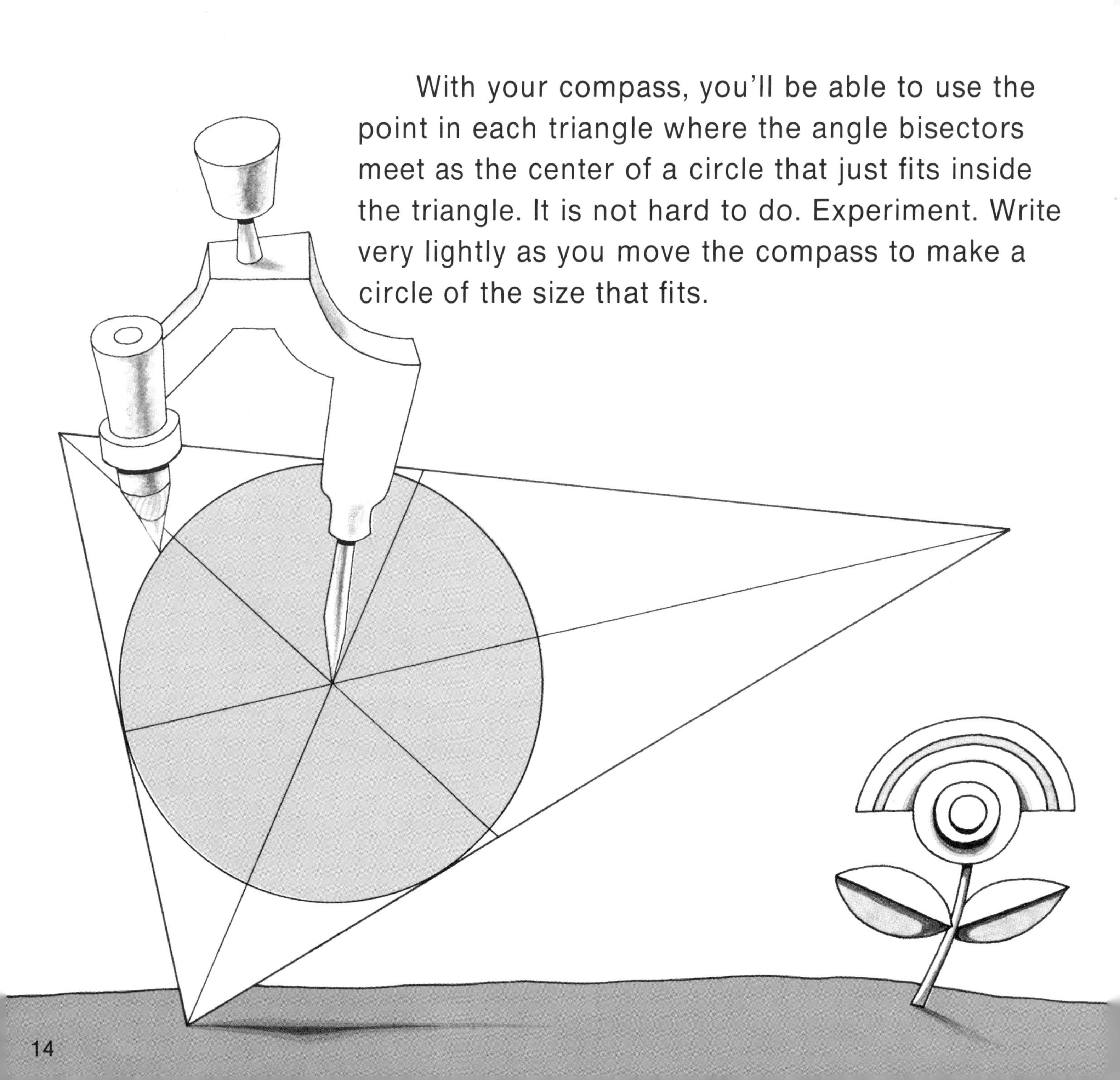

With your compass, you'll be able to use the point in each triangle where the angle bisectors meet as the center of a circle that just fits inside the triangle. It is not hard to do. Experiment. Write very lightly as you move the compass to make a circle of the size that fits.

The sides of a triangle are segments. A SEGMENT is a piece of straight line that has a starting point and a stopping point. A segment has two endpoints.

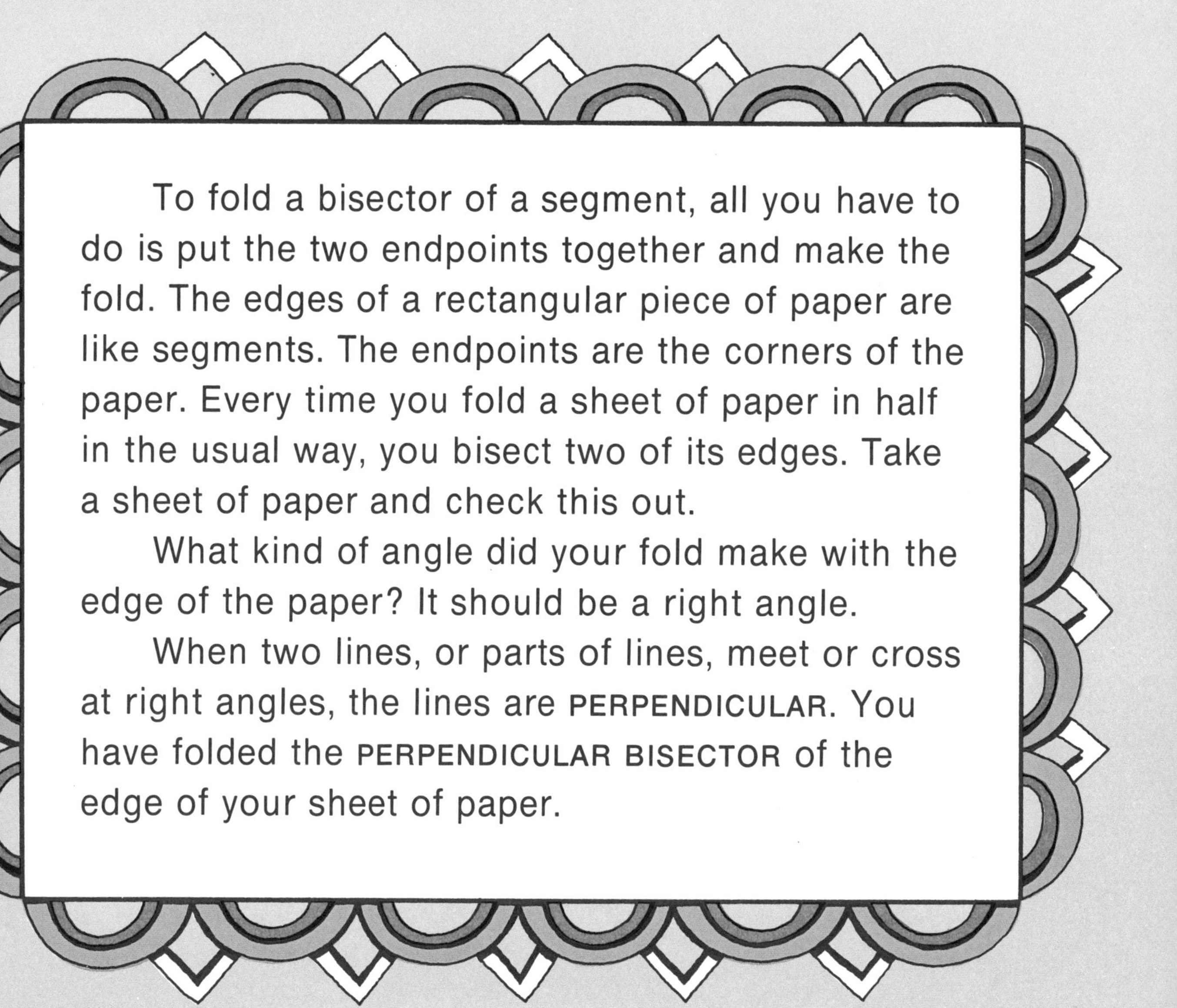

To fold a bisector of a segment, all you have to do is put the two endpoints together and make the fold. The edges of a rectangular piece of paper are like segments. The endpoints are the corners of the paper. Every time you fold a sheet of paper in half in the usual way, you bisect two of its edges. Take a sheet of paper and check this out.

What kind of angle did your fold make with the edge of the paper? It should be a right angle.

When two lines, or parts of lines, meet or cross at right angles, the lines are PERPENDICULAR. You have folded the PERPENDICULAR BISECTOR of the edge of your sheet of paper.

PERPENDICULAR BISECTOR

PERPENDICULAR BISECTOR

Draw one of the triangles on page 13 again. The vertex of each angle is also a vertex of the triangle. The plural of VERTEX is VERTICES. Put two of the vertices together and fold the perpendicular bisector of a side. It may take some practice for you to be able to do this well.

Make some new drawings of triangles like those on page 13. Fold the perpendicular bisectors of all three sides of each triangle. For the triangle with an angle bigger than a right angle, you must not cut out the triangle. If you do, you will not be able to see the result.

Hmm! The perpendicular bisectors of the sides of a triangle meet at one point, too. With your compass, you'll be able to use the point where the perpendicular bisectors of the sides meet as the center of a circle that just fits outside the triangle.

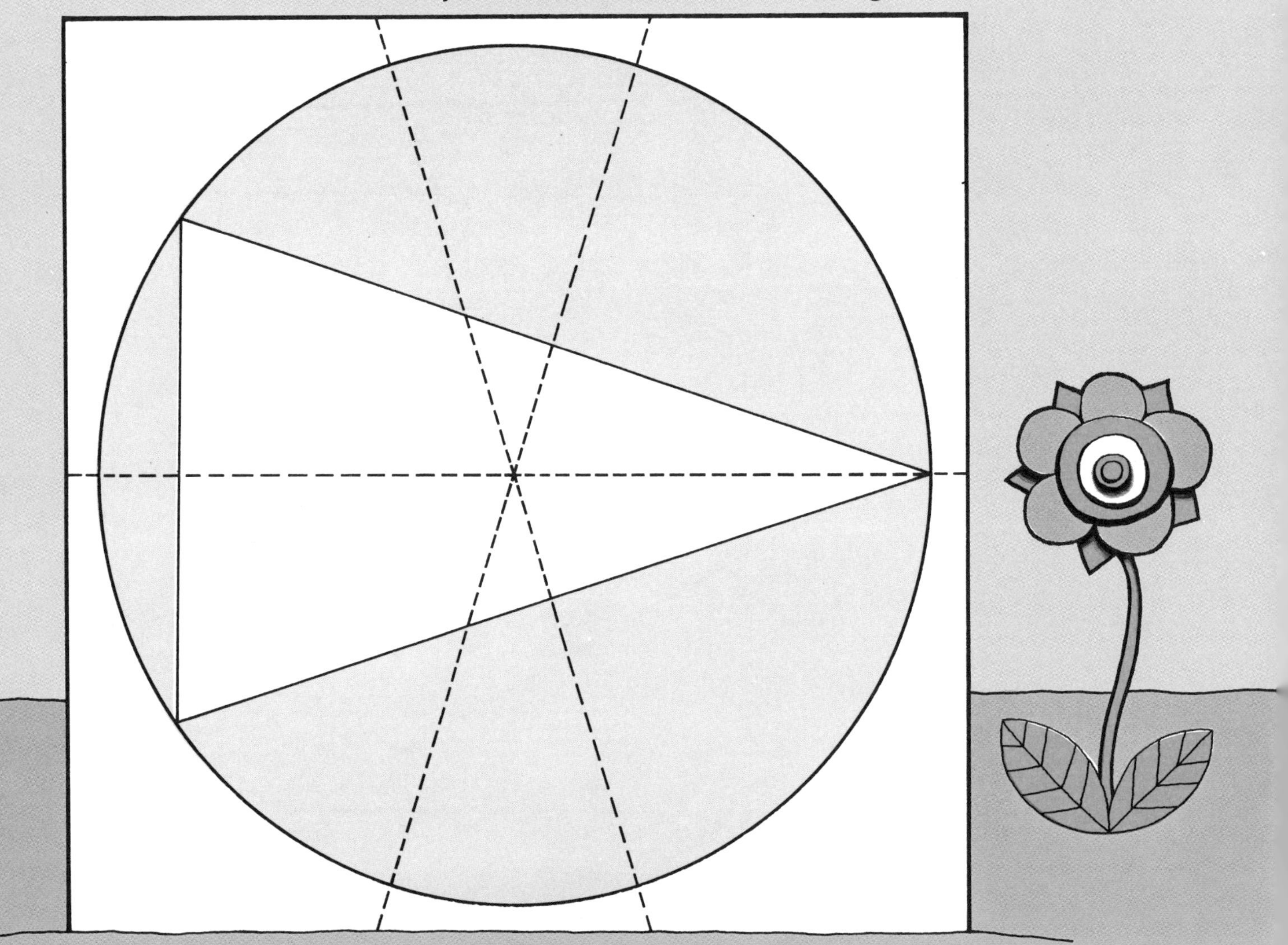

It goes through all three vertices. This will work regardless of whether the center point is inside the triangle, on the triangle, or outside the triangle.

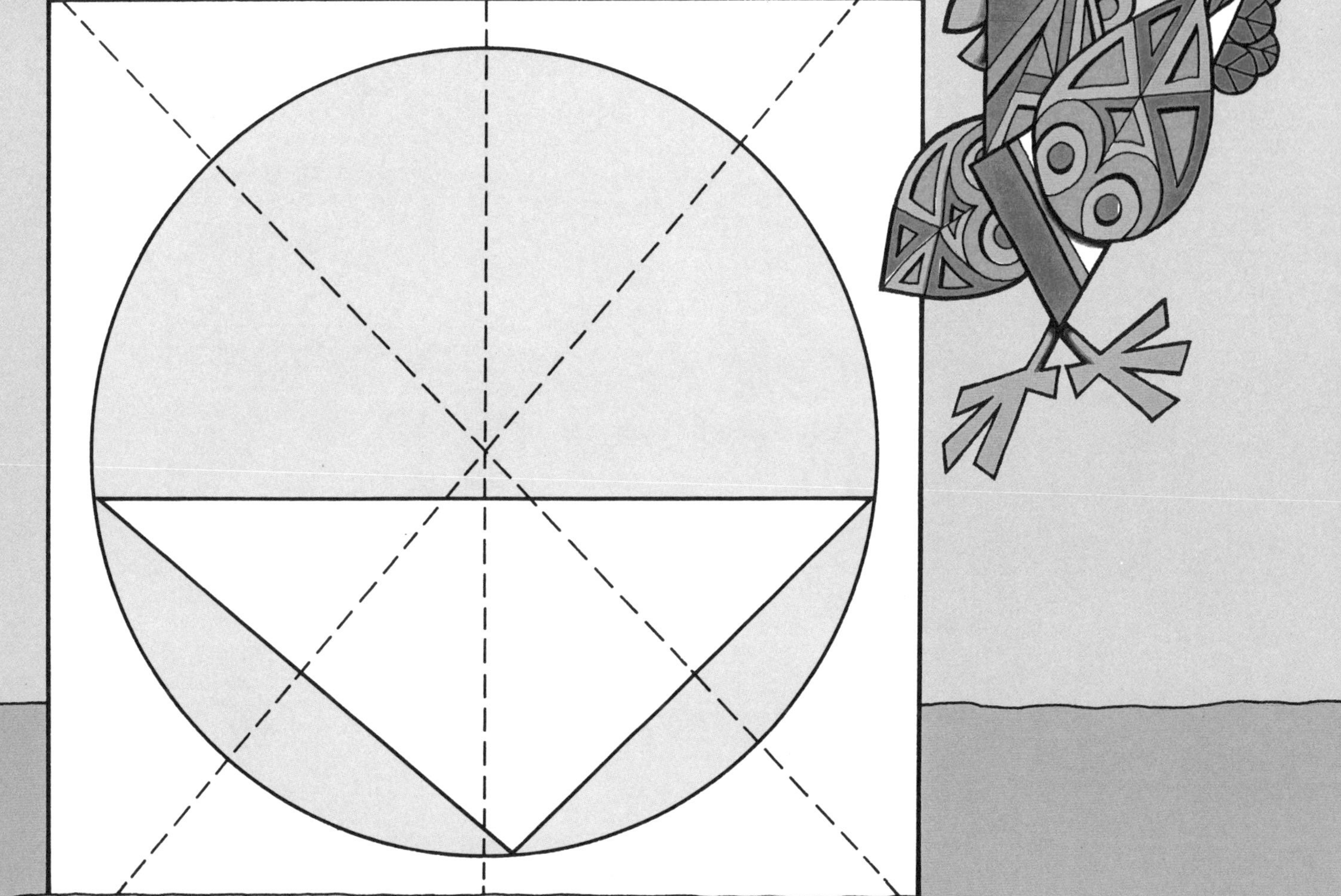

There is another kind of "middle line" for triangles. It is called a MEDIAN. Think what median usually means. Where is the median strip in a highway?

In a triangle a median is a segment that connects a vertex with the midpoint of the opposite side.

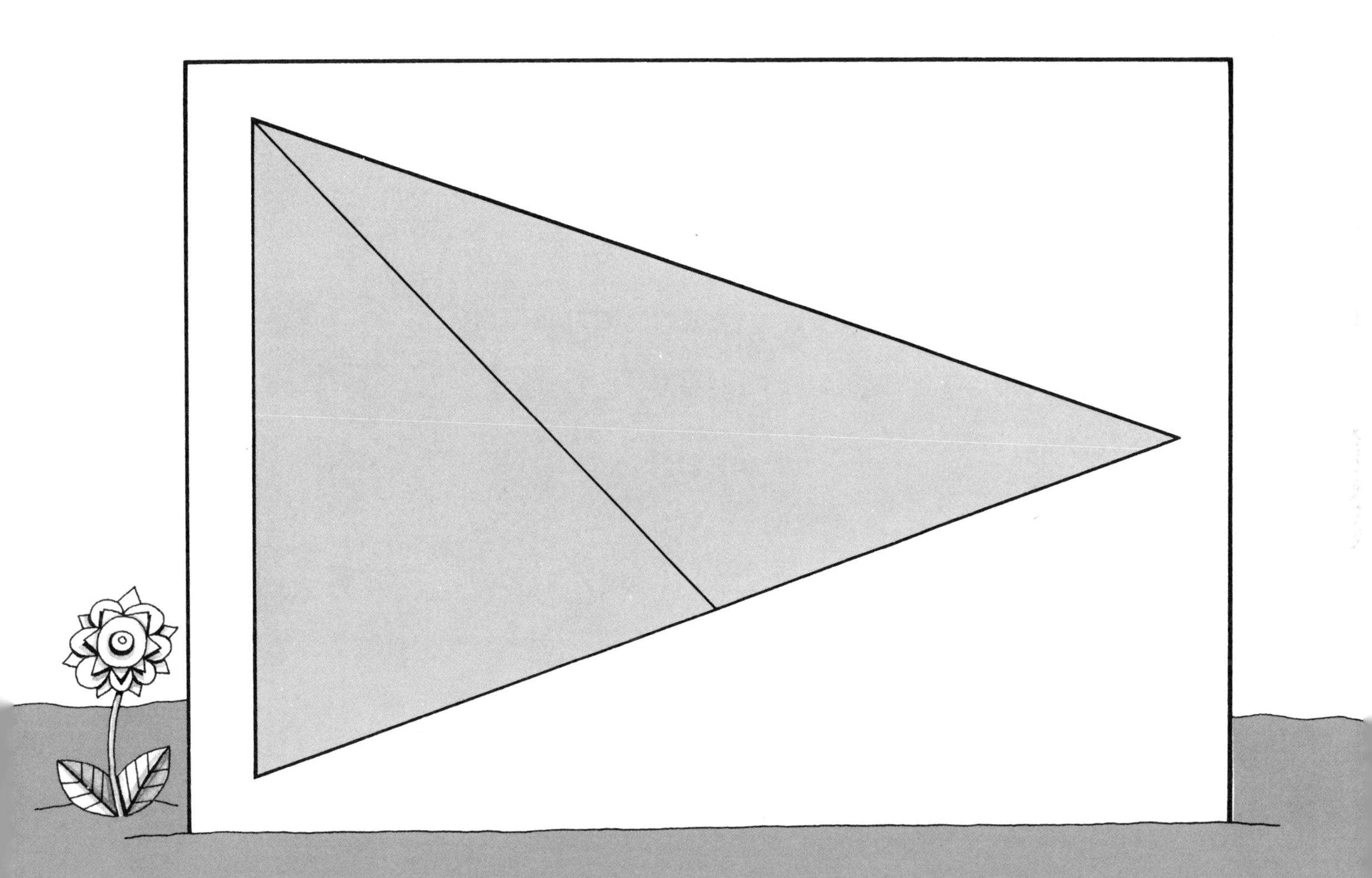

MIDPOINT means what it sounds like. It is the point that divides a segment in half—the point in the middle.

To find the midpoint of a side of a triangle, pretend you are going to fold the perpendicular bisector, but do not crease the fold. Just pinch the paper at the point where the fold would cross the segment, then open the paper and mark that point with your pencil. Take a piece of paper and your ruler, draw a couple of segments, and try this.

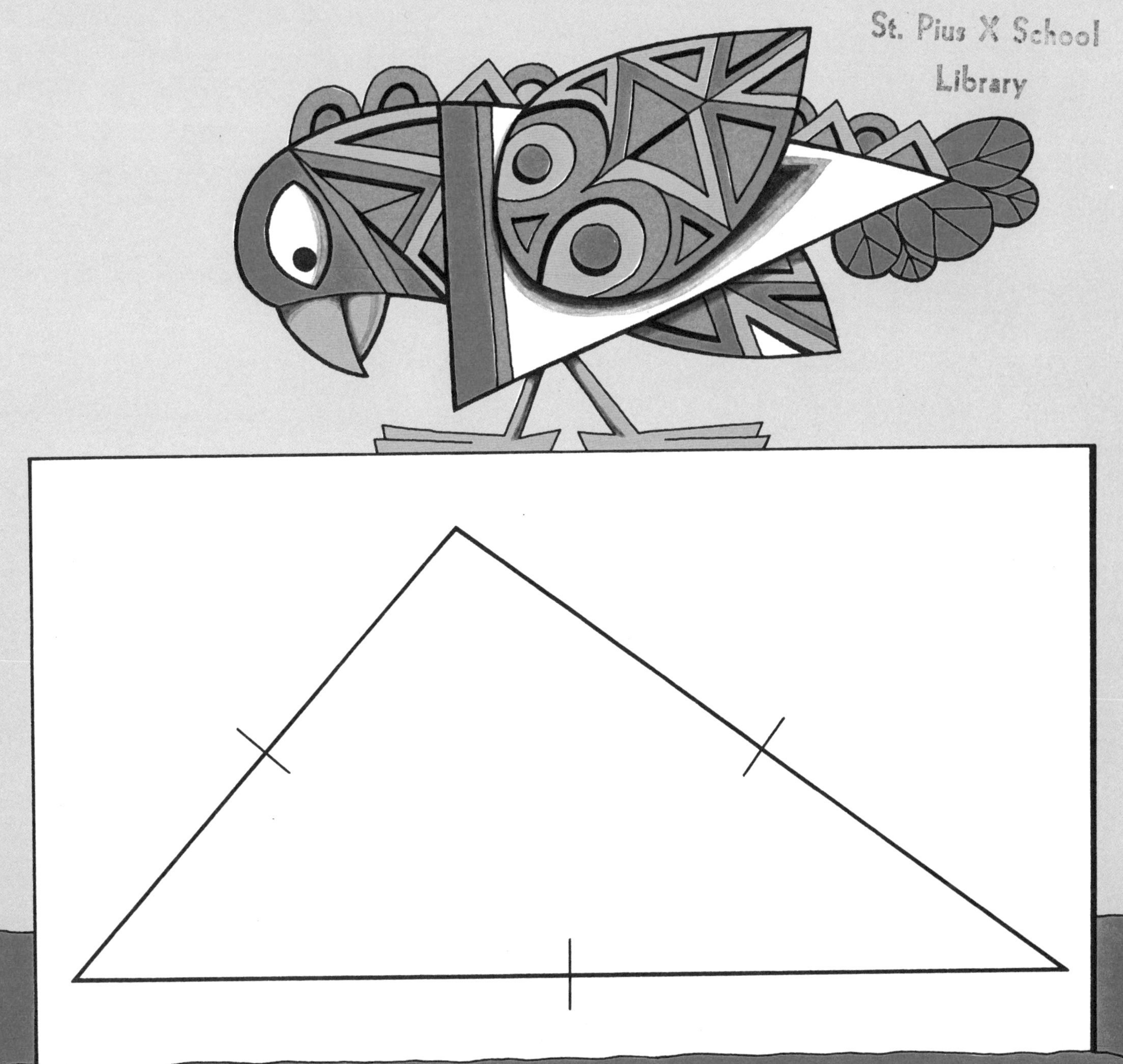

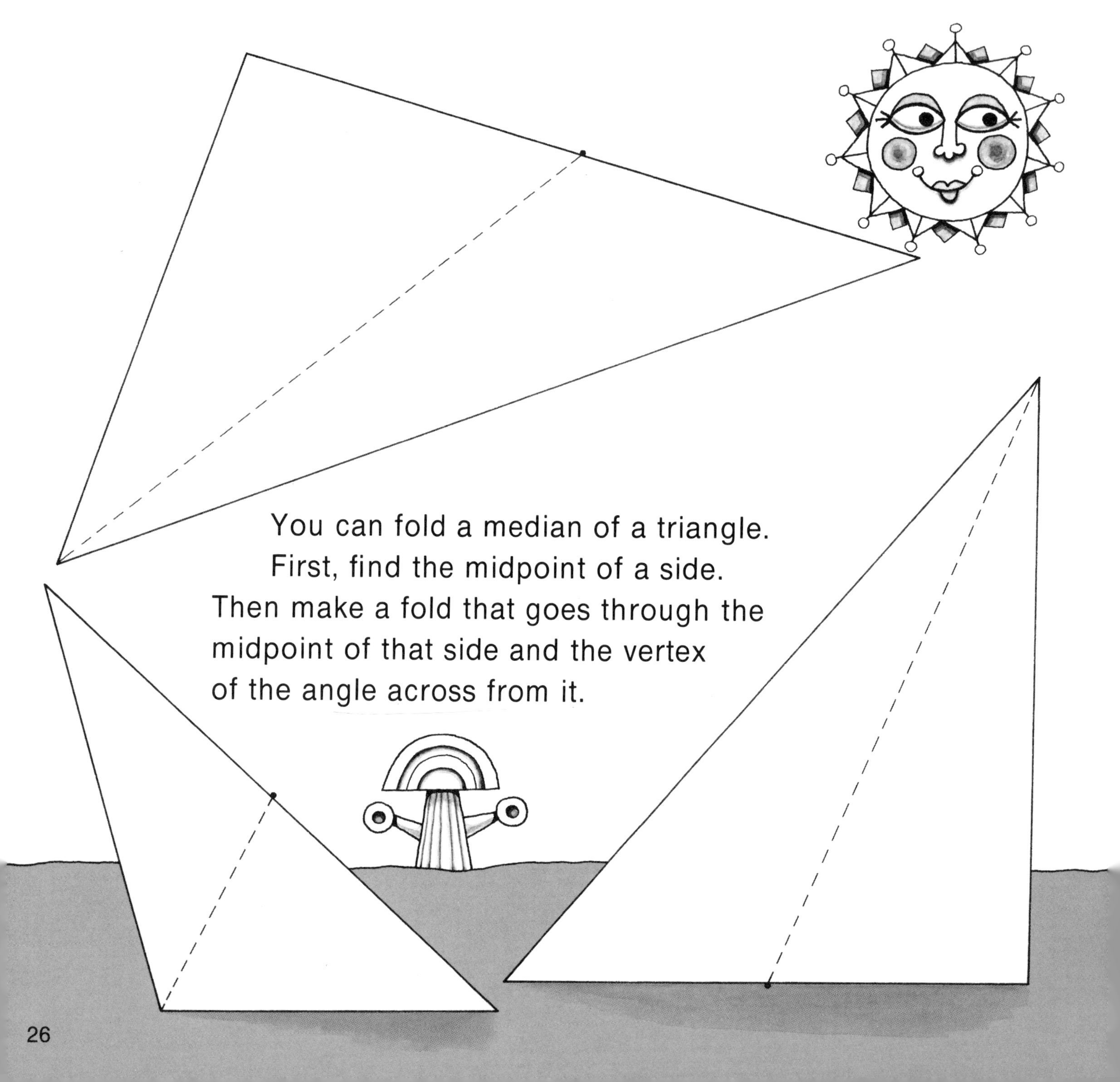

You can fold a median of a triangle.
First, find the midpoint of a side.
Then make a fold that goes through the midpoint of that side and the vertex of the angle across from it.

Practice folding medians so that each fold really goes through a vertex and the midpoint of a side.

Now make some new drawings of triangles like those on page 13, and fold all three medians of each triangle.

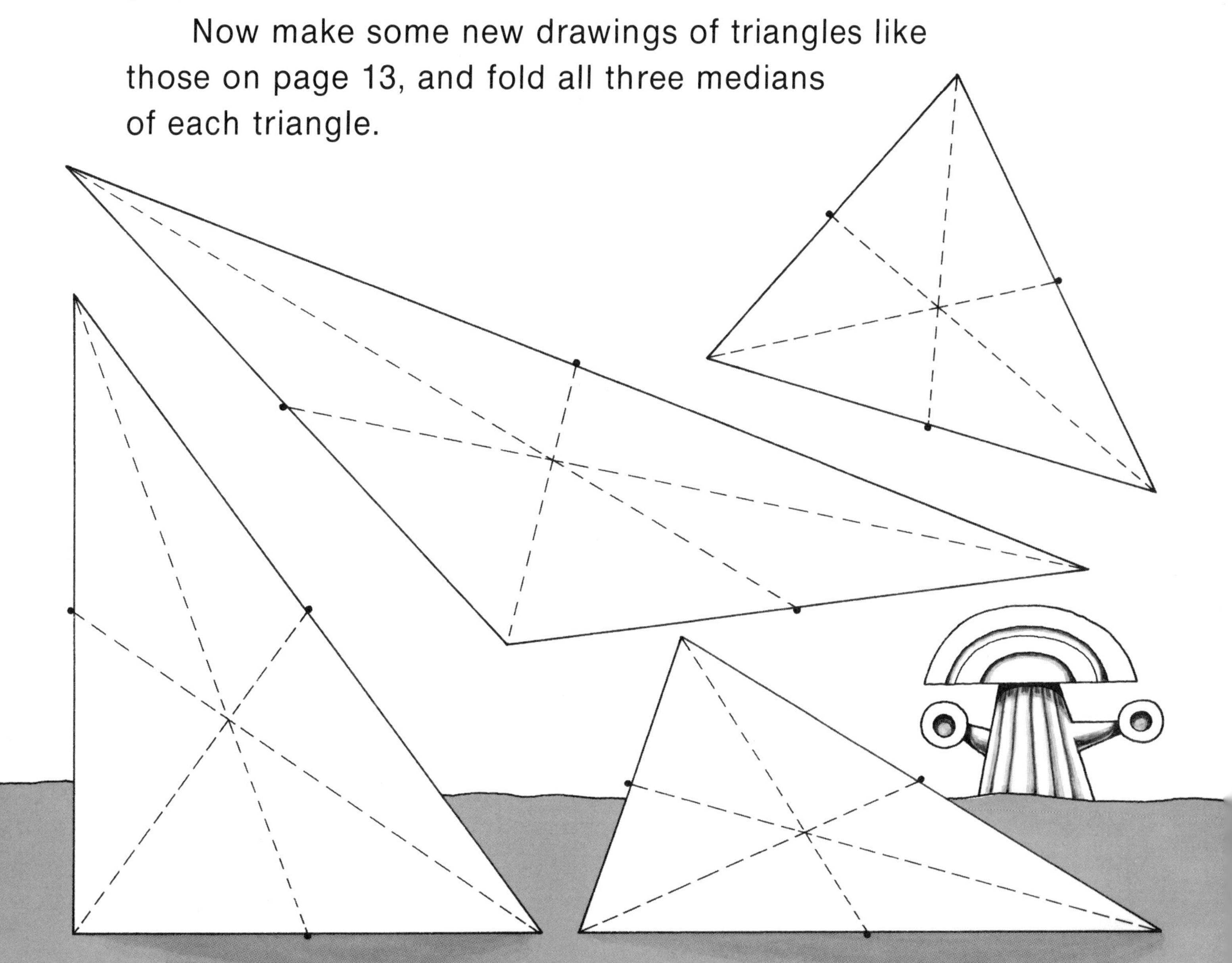

Were you surprised by what happened?

The three medians of a triangle meet in one point. The point where the medians meet is the CENTER OF GRAVITY of the triangle. The center of gravity is not the center of a special circle that goes with the triangle. A center of gravity is a sort of balance point. To see how this works out, get a piece of smooth cardboard and cut out a triangle. Find and mark the point where the medians meet.

To do this, trace your triangle on thin paper. Fold the medians. Put the tracing on top of the cardboard triangle so that all the parts match.

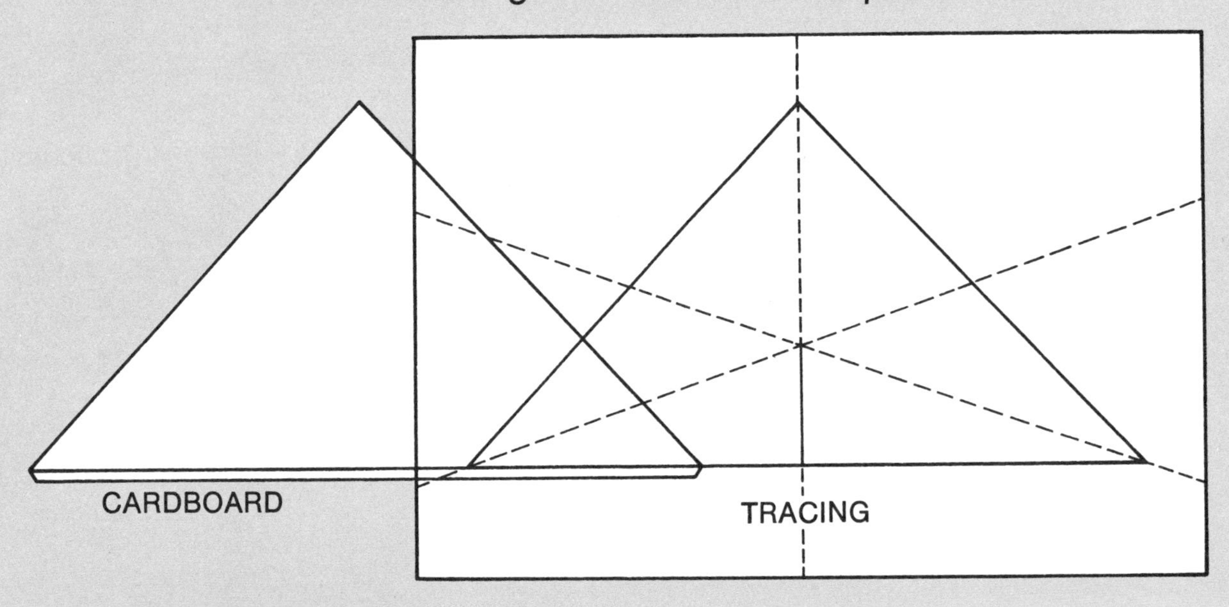

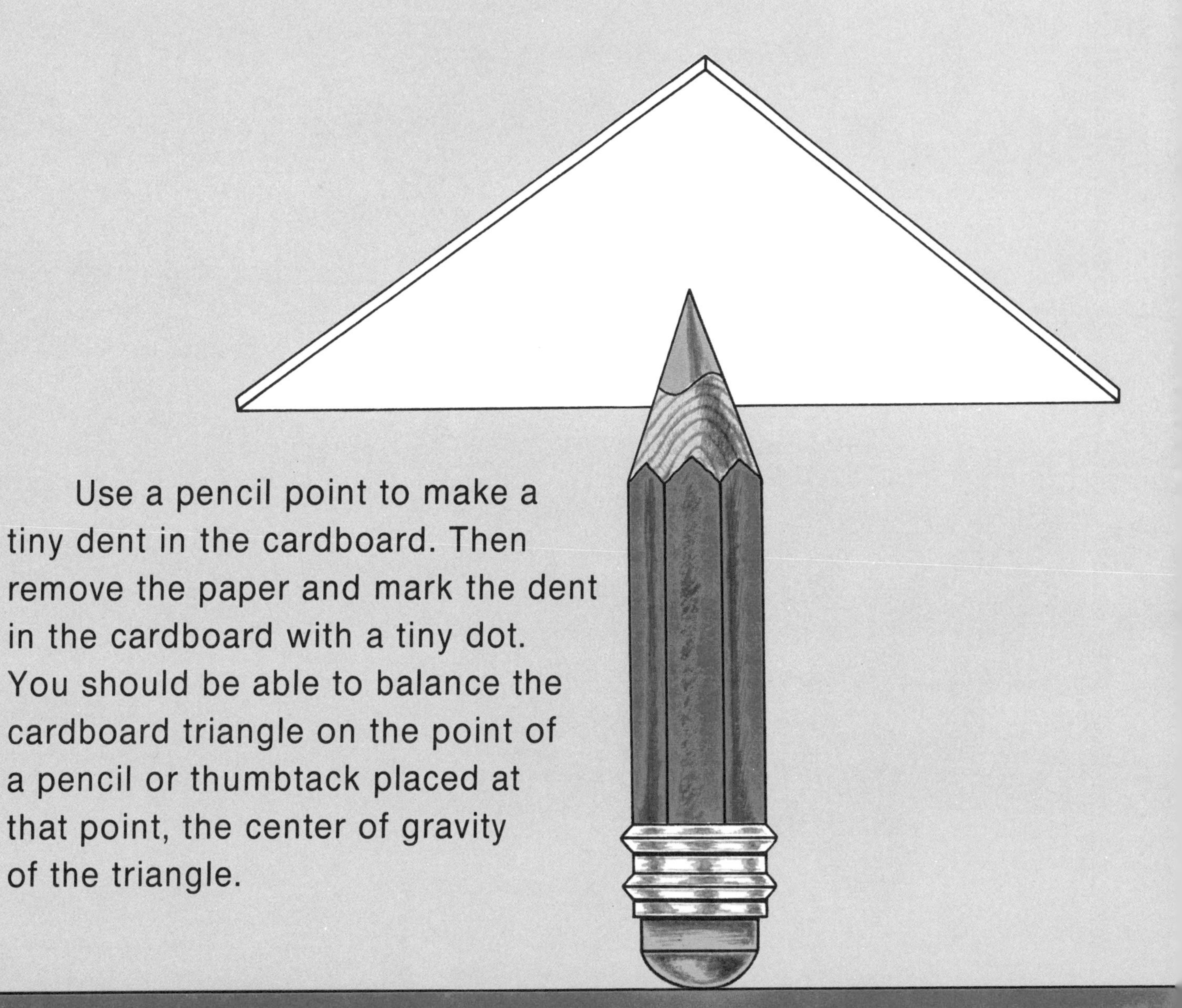

Use a pencil point to make a tiny dent in the cardboard. Then remove the paper and mark the dent in the cardboard with a tiny dot. You should be able to balance the cardboard triangle on the point of a pencil or thumbtack placed at that point, the center of gravity of the triangle.

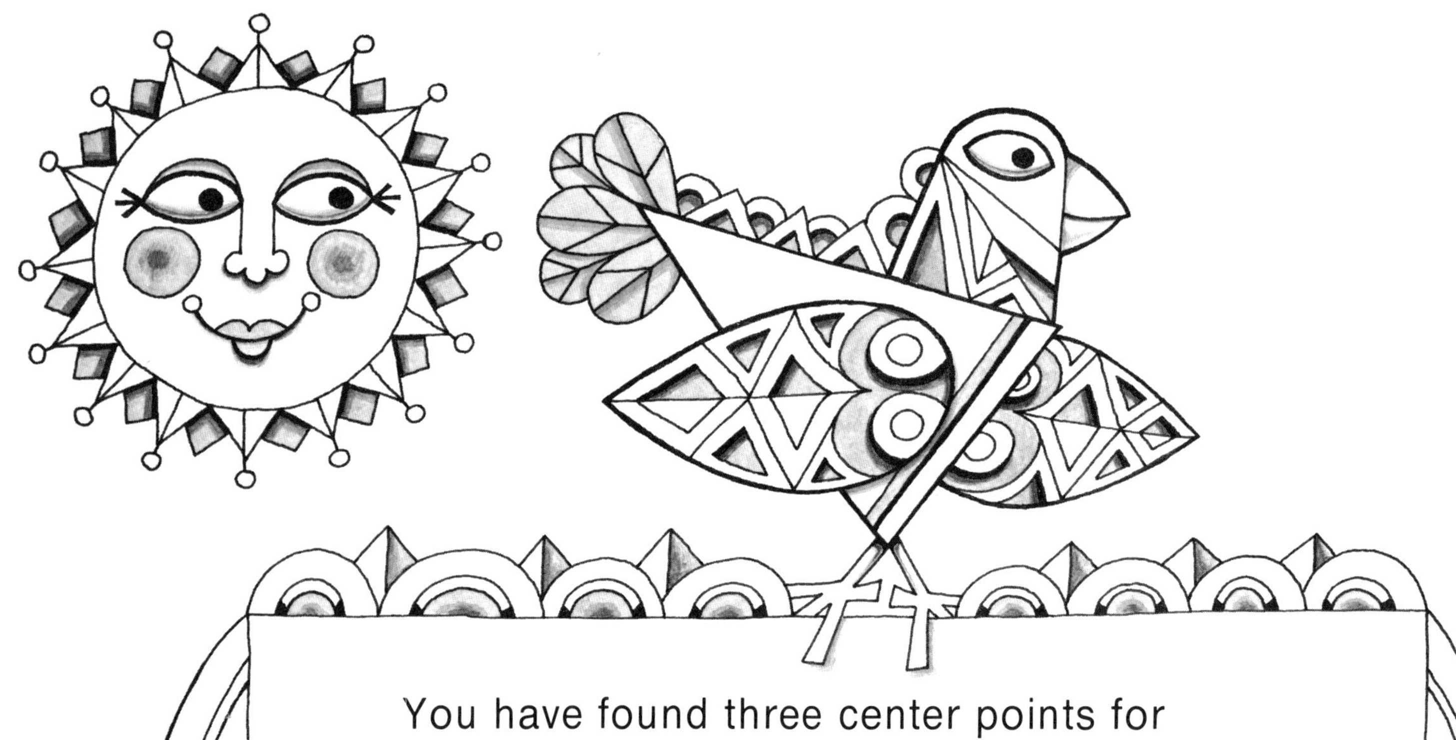

You have found three center points for triangles. If you have not already noticed something special about triangles that have all three sides the same length, make a drawing of the equilateral triangle on page 31. Fold the angle bisectors, the perpendicular bisectors of the sides, and the medians. It won't be much trouble. They're all the same! The center of the circle that fits inside the triangle is the same point as the center of the circle that fits outside the triangle and as the center of gravity.

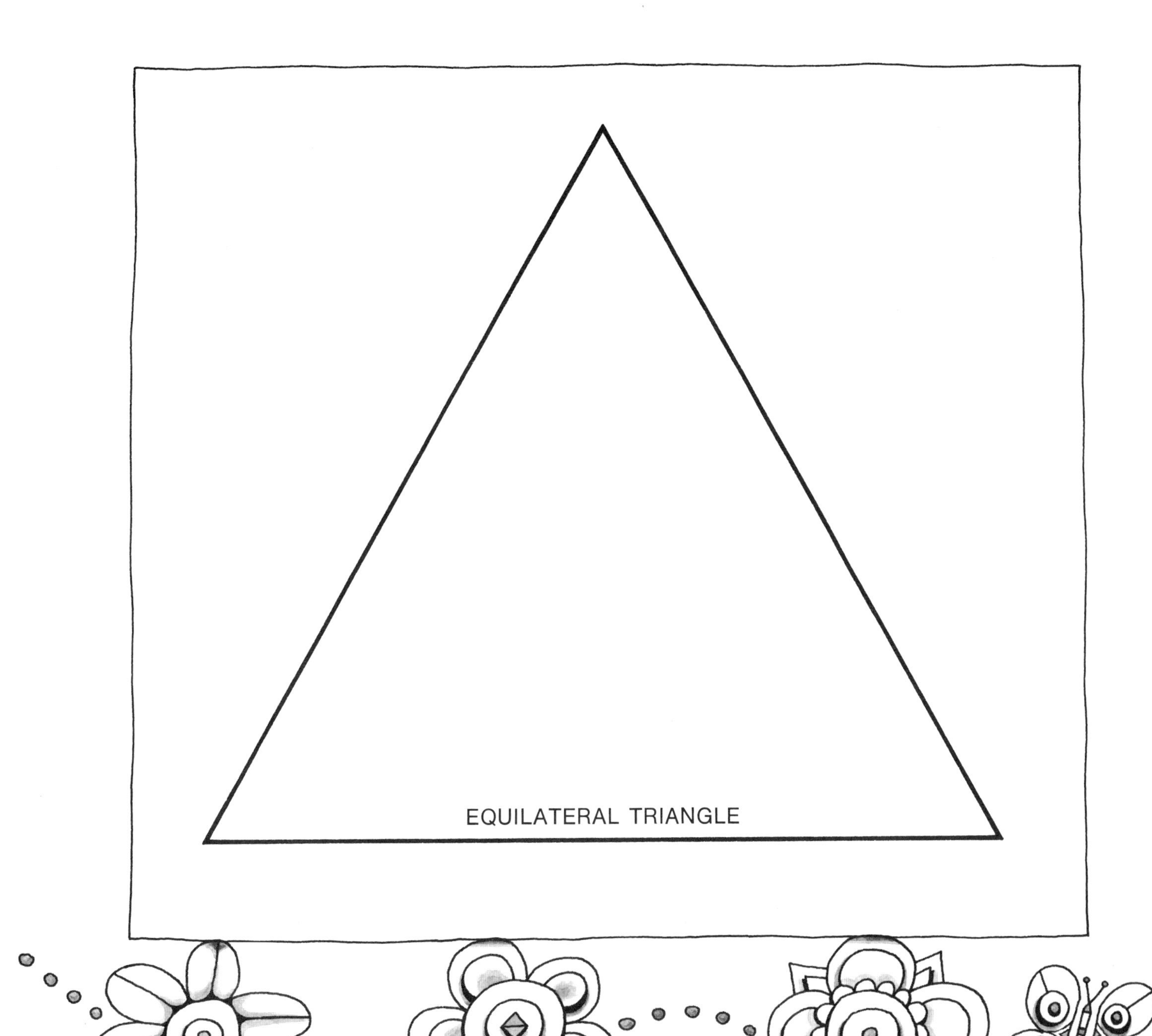
EQUILATERAL TRIANGLE

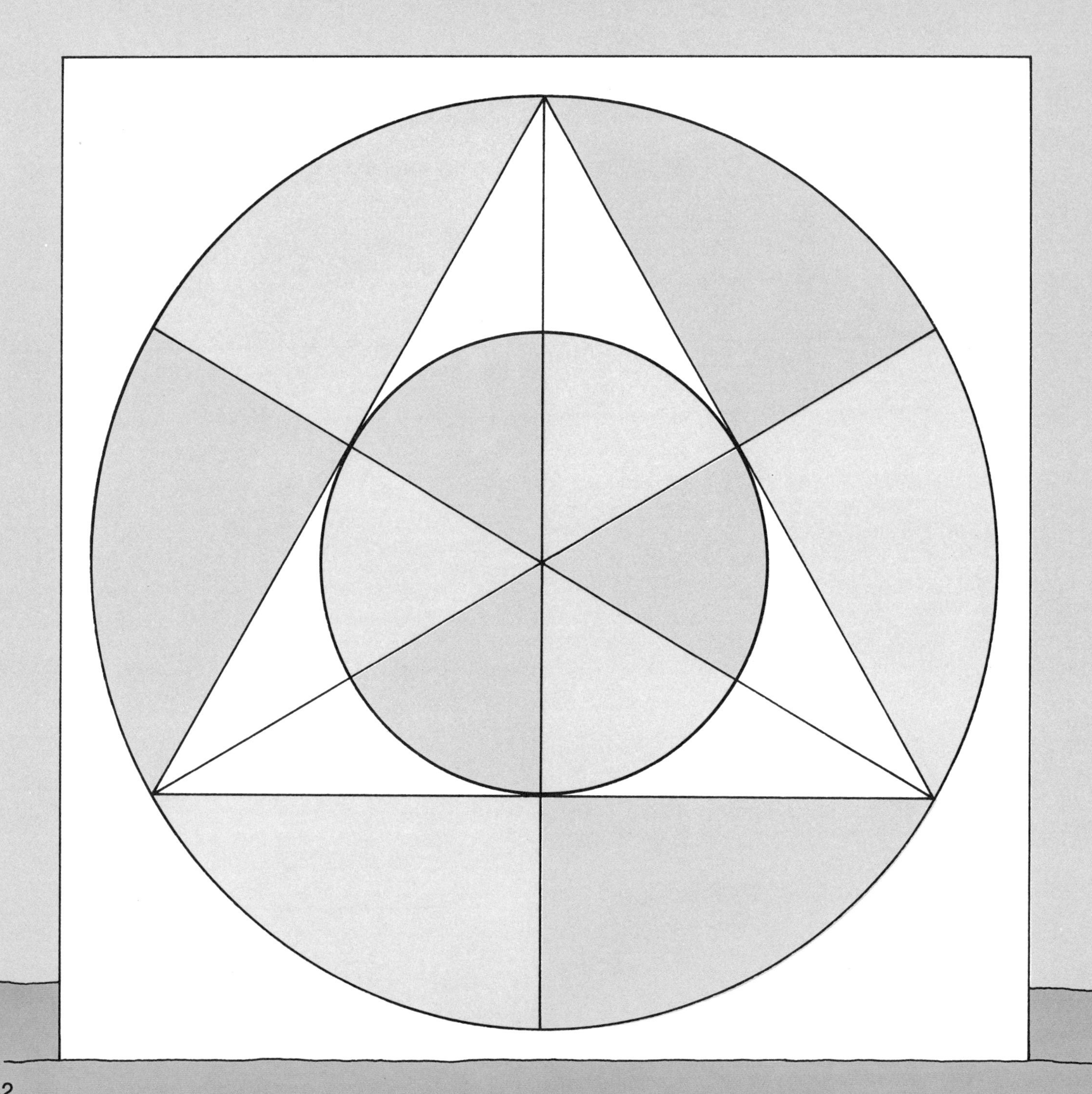

Triangles are wonderful. There are a lot more interesting things to learn about them. You have a good start now.

ABOUT THE AUTHOR

Jo Phillips believes that the best way to learn about math is to do projects and experiments. That is one reason why she enjoyed writing this book. She is also the author of *Right Angles: Paper-Folding Geometry,* another book in the Young Math series.

Dr. Phillips has taught school at every level from second grade through graduate school; she has also been a textbook author and editor, contributing editor in mathematics for *The Instructor* magazine, and an officer in the United States Coast Guard. She is now teaching teachers of mathematics at the University of Cincinnati.

ABOUT THE ILLUSTRATOR

Jim Rolling studied at New York's High School of Art and Design, the Art Students League, and Pratt Institute. He works in advertising and also designs type. One of his type faces appears on the jacket and title page of this book.

Mr. Rolling was born and raised in Brooklyn, and he now lives there with his wife and four children.

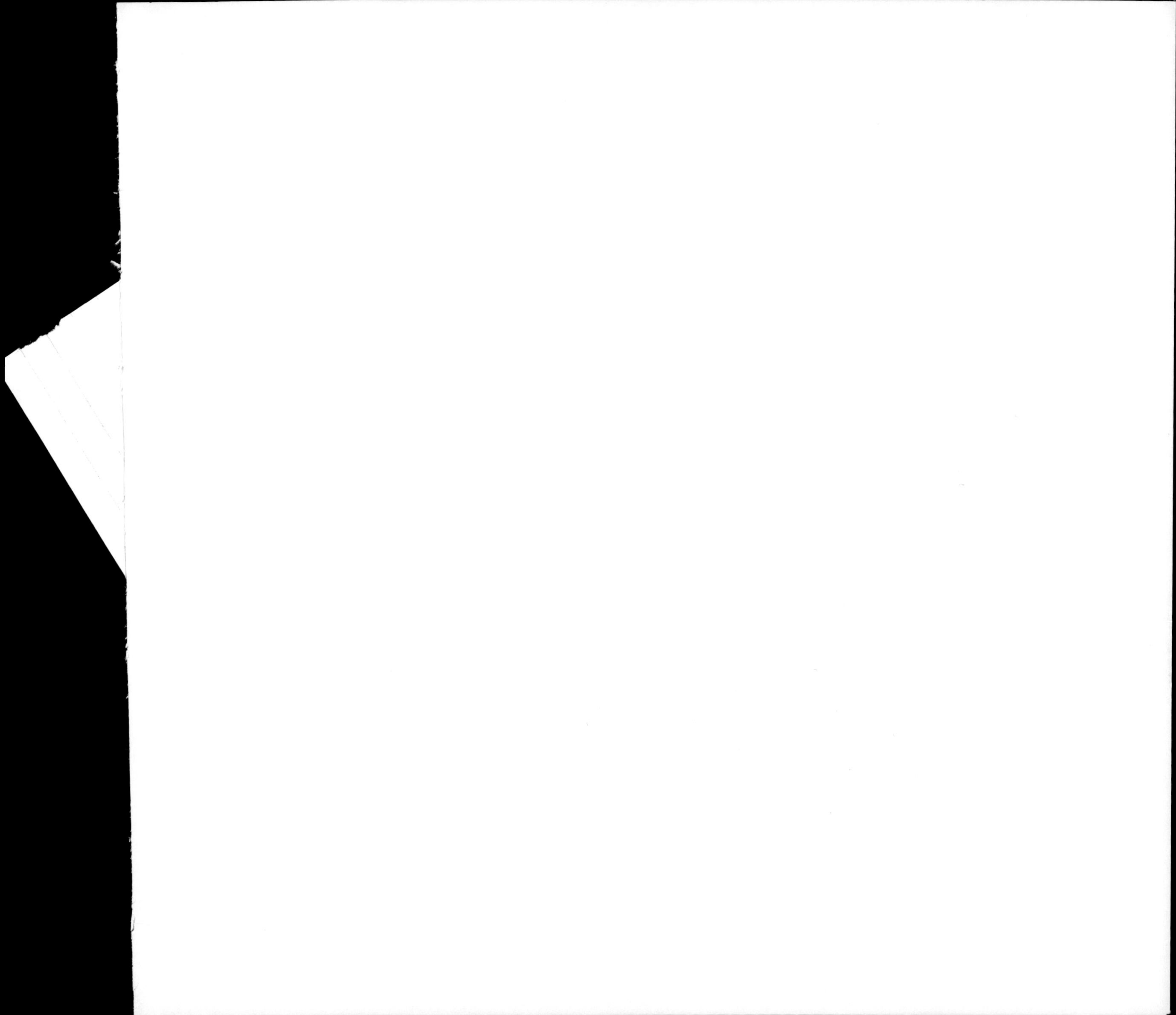

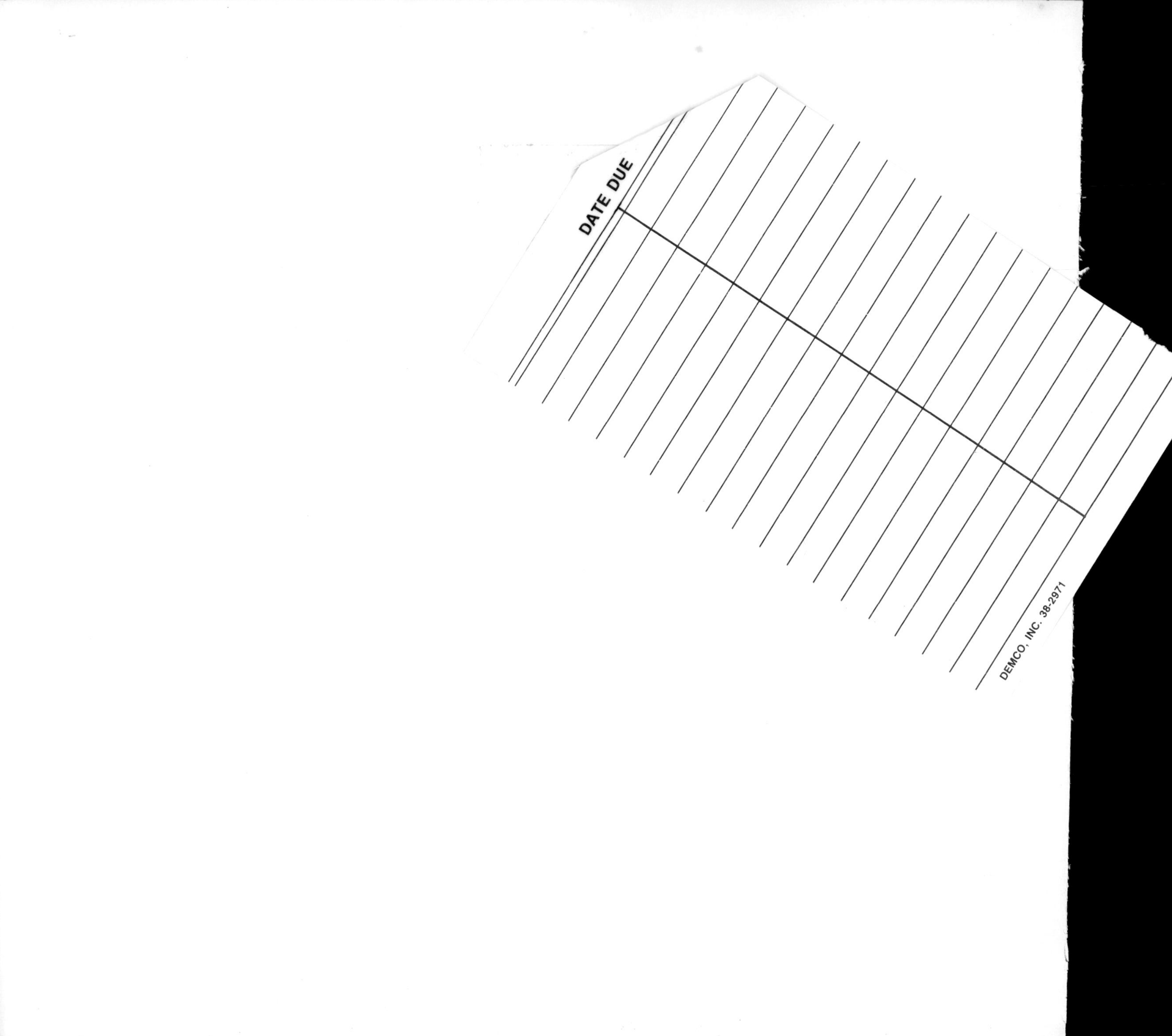
DATE DUE
DEMCO, INC. 38-2971